We find the state of affairs that binds us to our random
and ephemeral individuality hard to bear. Along with our tormenting desire
that this evanescent thing should last, there stands our obsession with
a primal continuity linking us with everything that is.

GEORGES BATAILLE

HALVOR NORDBY | ANDRÉ GALI

COSMIC DEBRIS
METEORITES AND JEWELLERY OBJECTS

BY REINHOLD ZIEGLER

ARNOLDSCHE ART PUBLISHERS

IMPRINT

© 2014 ARNOLDSCHE Art Publishers, Stuttgart, and the authors

All rights reserved. No part of this work may be reproduced or used in any forms or by any means (graphic, electronic or mechanical, including photocopying or information storage and retrieval systems) without written permission from the copyright holder ARNOLDSCHE Art Publishers, Liststraße 9, D-70180 Stuttgart.
www.arnoldsche.com

ARNOLDSCHE PROJECT COORDINATOR
Marion Boschka

TRANSLATIONS
Arlyne Moi, Stavanger (Norwegian | English)
Joan Clough, Castallack (German | English)
Cora Dannatt, Oslo (Norwegian | English)

LAYOUT
Silke Nalbach, nalbach typografik, Mannheim

OFFSET REPRODUCTIONS
Repromayer, Reutlingen

PRINTED BY
Leibfarth & Schwarz, Dettingen/Erms

PAPER
LuxoArt Samt 150 gsm

Printed on PEFC certified paper. This certificate stands throughout Europe for long-term sustainable forest management in a multi-stakeholder process.

Bibliographic information published by the Deutsche Nationalbibliothek
The Deutsche Nationalbibliothek lists this publication in the Deutsche Nationalbibliografie; detailed bibliographic data are available at www.d-nb.de.

ISBN 978-3-89790-405-7

Made in Germany, 2014

FRONTISPIECE
Ovoid Meteorite, 2013 (pp. 28/29)

This book has been produced with the generous support of

Norwegian Ministry of Foreign Affairs

Norwegian Association for Arts and Crafts

Arts Council Norway

CONTENTS

6   FOREWORD
Bernhard Wittenbrink

8   METEORITES

17   DISTANCE AND NEARNESS.
A Philosophical Essay on Primary Qualities in Art Objects

17   HORIZONS OF UNDERSTANDING
31   BEYOND CONCEPTS
40   THEORY OF CAUSATION
46   NORMATIVE ASPECTS

Halvor Nordby

52   GRAVITY

63   ON IDENTITY IN ART AND LIFE.
A Conversation Between André Gali and Reinhold Ziegler

72   TOOLS

79   Appendix

FOREWORD

Diamonds are overrated.

He did not use a hammer to pound in nails. Thor, the God of Thunder, hurled a hammer into the world.

In the same way, the hammer designed by Reinhold Ziegler as a jewellery object has also sought a new function. The jewellery object soon found it with a new owner. Not as a hammer, which it could not be used as, but as a communicator between its owner and his or her environment. Thus the jewellery object has found its cosmos and at the same time is for Reinhold Ziegler like an outpost, a satellite, which reports from and on Reinhold Ziegler.

So Reinhold Ziegler has also thrown his hammer into the world.

The next step. His most recent works take back the material from the world. A piece of rock was hit and smashed by Thor's hammer billions of years ago and meteorites rain down on earth.

A material billions of years old is the basis for Reinhold Ziegler's new works.

A meteorite with its austere beauty becomes a new ambassador, hence a communicator, for the woman or man who wears these jewellery objects. A meteoroid burns out as a shooting star in the earth's atmosphere but reaches the ground as a meteorite. When worn, it does not sparkle like a diamond, whose story is so often summed up with 'cut', 'beautiful' and 'expensive'; instead it tells the infinite tale of our world and our time from the beginning. And it did sparkle in outer space. Meteorites contain the most ancient material in our solar system and came into being 4.5 billion years ago. They provide the sole direct access to the genesis of the solar system in its ineffable beauty. When a meteorite is cut into by Reinhold Ziegler, crystalline structures are revealed with a beauty that only appears in meteorites. Those adamantine structures are created while meteorites are cooling down very slowly over millions of years. An inner sparkle.

So one might say diamonds are overrated although a meteorite also conjures up diamonds: micro-diamonds are created by the impact of a meteorite.

Be that as it may … since there is a consensus on the value of diamonds, jewellery and the jewellery industry nowadays centre on that valuation. Nonetheless, the true teller of tales among jewellery objects that link us with the world and time is the meteorite. Even though it is deployed once again by Reinhold Ziegler as a tool without a function.

Bernhard Wittenbrink

WEATHERED FUSION CRUSTED METEORITE – pendant – 2013 – stony meteorite NWA (Unclassified, North West Africa), silver, nylon cord – 42 × 68 × 33 mm

METEORITE BALL – pendant – 2013 – stony meteorite NWA (Unclassified, North West Africa), silver, leather cord – 60 × 54 × 47 mm

Reinhold Ziegler wearing
METEORITE CONE

Morten Bilet wearing
METEORITE BALL

METEORITE CONE – pendant – 2013 – stony meteorite NWA (Unclassified, North West Africa), silver, nylon cord – 80×35 mm

Sigurd Bronger wearing
WIDMANSTÄTTEN STRUCTURED
METEORITE 1

DISTANCE AND NEARNESS
A PHILOSOPHICAL ESSAY ON PRIMARY QUALITIES IN ART OBJECTS

Halvor Nordby

→ HORIZONS OF UNDERSTANDING[1]

The jewellery objects of Reinhold Ziegler express a basic tension between distance and nearness. They are distanced in the sense that they consist of elements that represent and refer to things in an external world about which we have limited knowledge. Meteorites – the theme of *Cosmic Debris* – usually come from the so-called asteroid belt that lies between the orbits of Mars and Jupiter. Their origin, about 4,57 billion years ago (one billion years older than our own planet), and journey through outer space exceed the boundaries of our physical and rational horizon of understanding. Meteorites are coagulated stardust from the time when our solar system was formed. They come from the same material that provides the basis for our existence. But for us, the origin of stardust and the solar system are distant realities we cannot access through direct observation. Meteorites take us closer to the origin of being. They are extra-terrestrial bodies captured by the earth's gravitation, but they also implicate something about our own origin when they become part of our experienced reality here on this planet.

In this essay, I reflect on what it means to approach these works via concepts, especially concepts integral to the field of philosophical aesthetics – or directly, *sans langage*, through the senses. I also inquire into the limits of human knowledge by discussing what it means to predicate primary and secondary qualities to these works, and by elucidating what they can imply about viewing ourselves from a first-person or a third-person perspective. My aim is to show some of the depth and breadth of meaning *Cosmic Debris* affords about the position of humanity in a wider horizon of understanding.[2]

→ INTERPRETIVE PERSPECTIVES

Generally speaking, the jewellery objects in *Cosmic Debris* are pure and fundamental in form. They create sensory impressions of something basic and genuine, which exists in, or at the very least is constituted by, the natural reality around us. The meteorite works draw associations to ontological realism: the idea that objects in the external world are as they are independent of how we experience them (Stroud 2000). Ziegler is preoccupied with external and authentic aspects of the world that are minimally affected by the viewer's standpoint and need to categorize.

In contrast to a traditional individualistic jewellery object that is meant to accentuate the wearer, it is now the wearer who should accentuate the jewellery object. He or she stands as a facilitator and then recedes into the background. The wearer is drawn into something essential, beautiful and remote, which comes from another place. Also the time the wearer is allotted here on earth is reduced and placed in a more comprehensive interpretive frame. The experience of standing before a larger, eternal universe of meteorites and interpretive spaces is accentuated and becomes dominating. The inspiration arises to see oneself from the outside.

One of the jewellery objects, *Meteorite Ball* (p. 10), brings to mind a globe, since it has a ring-shaped holder similar to the gadget a globe is fastened in, and which enables it to spin in all directions. The polished meteorite was not initially part of our planet, and as a representative of the concept *meteorite,* it still does not have a natural existence here on earth. Precisely for this reason, it becomes a representation for us. A representation can never be identical to what it refers to – it must be something other than the object of reference. But the meteorite is a *natural* representation. It is a natural substance with natural properties.

At the same time, all the jewellery objects in *Cosmic Debris* are torn from the natural universe, in the sense that the physical material is absolutely affected and shaped – in the first instance by the artist. These works are also physically influenced, and *should* be influenced, by those who wear them. Many of them can be worn around the neck, and the mechanisms to which they are fastened emphasize their functionality. The meteorite parts can be turned in several directions, relatively freely, as in *Meteorite Chisel* (pp. 34/35), where the fastener is formed like a carabiner, or more limitedly, as in *Meteorite Cone* (p. 15), where the fastener is a Cardan (u-joint) mechanism. These fasteners are manmade and give us as wearers the opportunity to move them about in any direction. Anyone who appears as a *possessor* of one of these meteorites has complete control and can carry out repetitive actions. The jewellery objects, as representations of something larger than ourselves, can be controlled as systems, but the controlling movements are not ordinary activities. We carry out a pattern of action that creates distance to everyday routines.

→ CONCEPTS

We do not, however, have conceptual power over the jewellery objects. We cannot reduce their aesthetic to concepts and descriptions. In fact, on several occasions Ziegler has stressed that he does not consider himself a pure conceptualist. His goal is not primarily to convey ideas through artistic expressions, as has been the case in conceptual art.

The marked distance from purely conceptual jewellery art comes clearly to the fore in *Cosmic Debris*. It is in the project's nature to point beyond language – it exceeds what we can reductively acknowledge and understand through concepts and ideas. It is not our need for explanation, or even our possibility of doing so, that should be satisfied. The goal is rather to show, from a cosmic perspective, how small and insignificant we actually are, and how humble and non-self-centred we should be when trying to understand ourselves in relation to a historical, all-encompassing horizon. When the means for achieving this are jewellery objects – objects that traditionally emphasize us as individual wearers – the distance to egocentric perspectives increases. Ziegler uses the concept of a piece of jewellery dialectically, to elucidate perspectives that are incommensurable with traditional individualistic ideas about jewellery. The meteorite works are jewellery objects, but they can also be understood as aesthetic antitheses of person-oriented art jewellery. Understood in this way, they appear as powerful analytical means for knowledge recognition (*Erkenntnis*).

Several of the jewellery objects underscore how limited our individual perspectives are. In *Widmanstätten Structured Meteorite* (p. 19) and *Square Meteorite in a Circle* (p. 49), we see a natural pattern of small crisscrossing lines and fields. The pattern is the result of an extremely slow cooling process – about one degree per 200,000 years – through which the minerals kamacite and taenite slowly exsolve (separate from each other at a critical point in temperature) deep within the extra-terrestrial body from which the meteorite originates (Smith, Russell and Benedix 2009). This we can see when a cross-section of the meteorite is etched with weak acid. The result is the so-called Widmanstätten pattern, named after the Austrian scientist who presumably first discovered it in 1808. Scientists

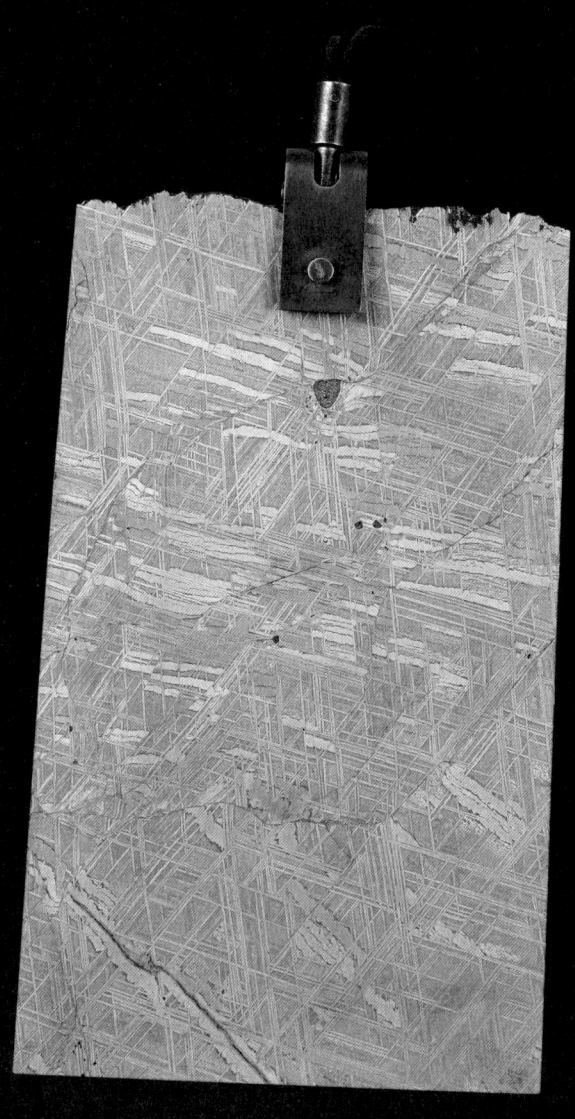

WIDMANSTÄTTEN STRUCTURED METEORITE 1
pendant – 2013 – iron meteorite Muonionalusta (Norbotten, Sweden), silver, leather cord – 153 × 80 × 10 mm

WIDMANSTÄTTEN STRUCTURED METEORITE 2
pendant – 2013 – iron meteorite Gibeon (Namaland, Namibia), silver, leather cord – 125×80×10 mm

study the patterns of meteorites because they indicate something about slow chemical processes which we have limited knowledge of, due to our own time frame. These patterns have not been observed in any material originating in our own earth.

Knowledge about the Widmannstätten pattern is something we can gain, and when we have it, it can help us see ourselves in a larger cosmic perspective. The realization that we stand before the evidence of a process that is much more comprehensive than our idiosyncratic thought processes – our personal interpretive perspective – causes the subjective understandings we have to appear random and insignificant by comparison. Our normative judgements of how we *should* understand the jewellery objects in *Cosmic Debris* no longer seem objectively true and profound. The judgements, proposed at a given point in time, are reduced to random, fleeting remarks in a little place, on a little planet, in an infinite universe. When we try to conceptualize – to codify the aesthetic content – our interpretations seem almost negligible in the eternal perspective.

At the same time, there is no sharp division between conceptual and non-conceptual art, and Ziegler's jewellery objects are necessarily affected conceptually by all who actively experience them. Experience presupposes interpretation, and concepts such as *jewellery, art jewellery,* or just *art*, are relevant and requisite for understanding. But these are man-made concepts, and we use them as grasping tools. As with all concepts, the understandings we take for granted are explications we develop ourselves. Our conceptual definitions are fictive and products of culture. From an interpreter's perspective in space and time, all concepts have a specific intention (meaning) and extension (reference in the world).

Concepts are *stereotypes* (Nolen-Hoeksema et al. 2009). They set conditions for naming objects and events. They divide the world into categories. Like pieces in a puzzle, we want them to come together into a unified and intelligible picture precisely when we want to understand. Based on our limited intellectual capacities, we seek the sort of understanding we can put into words, with concepts we possess and think we master (Nordby 2004).

This is an intuitive point from the philosophy of mind and language. But it is less obvious *how* objects exist independent of concept applications. When we stand before the jewellery objects in *Cosmic Debris*, the relation might appear uncomplicated: we are here and the objects are there, such as we experience them. We are observers, and the meteorite works apparently have a number of qualities that continue to exist even when we no longer pay attention to them. It seems obvious that whoever wears or interprets Ziegler's jewellery objects surely cannot change their *essential* inherent qualities.

→ PRIMARY AND SECONDARY QUALITIES

But it is not that easy. Take for instance something as simple as colour. We can easily imagine that the colours of a jewellery object – say of part of a meteorite – will continue to be there even when we no longer view them. Colours seem to be 'real' qualities we refer to when we say, for instance, that something is brown; it 'seems to be there', independent of our experience. But as the philosopher John Locke (1690) famously pointed out, colour, in reality, is something that emerges for each of us individually (Mackie 1976, Stroud 2000). Colour-blind people (also animals equipped with non-human sensory apparatuses) have different colour experiences than normally-sighted people.

TRANSLUCENT METEORITE – brooch – 2013 – stony-iron meteorite Seymchan (Magadan, Russia), silver, ebony, steel – 68 × 68 × 14 mm

Halvor Nordby wearing
TRANSLUCENT METEORITE

Colours and colour concepts are experiential – they are created through observation.

In *Cosmic Debris*, this principle about inherent qualities gains two additional points of significance. First, in works such as *Weathered Fusion Crusted Meteorite* (p. 9), the brown fusion crust of the meteorite is highlighted. It was created when the meteorite entered the earth's *atmosphere* – at a speed of between 40,000 – 250,000 km per hour – and the friction between the rock's surface and the atmosphere generated such enormous heat that the *meteorite's* surface melted and turned black. Over time, as the meteorite weathered on the ground and the iron in it rusted, it turned brown. The brown colour has thus been created by terrestrialization, by being on our planet. But the direct visual perception of the fusion crust as part of an artistic expression is something we cannot have until we see the object.

Second, we cannot be certain that the fusion crust appears the same way to every observer. Like other colours, it is logically possible that we all perceive it in different ways. Even here and now we lose objective control over the jewellery object – we do not even know if we are talking about the same thing. As long as we all use the same words in our evaluations, the differences in our colour perception will go unnoticed. Colour arises as an inner sensory experience; it is something to which others only have indirect access. As Stroud (2000, p. 72) observes, the conventional idea is that colours are secondary qualities arising through a sense relation between an object and a person who experiences the object:

What reveals that colour and other secondary qualities are mere appearances and are not part of the reality is that ... objects do not have to be thought of as having such qualities to explain why the world appears to all of us as it does.

The idea is that colours are aspects we notice on objects that exist 'out there'. Primary qualities stand in contrast to secondary qualities; they are qualities an object has independent of sense experience (Locke 1690, Mackie 1976). Primary qualities can be defined as the negation of secondary qualities – as qualities that *do not* arise as sense relations between objects and perceivers.

The primary qualities of the meteorites in *Cosmic Debris* existed before they were observed by us, and they will continue to exist just as they are as long as the objects do not change. They continue to be there when we, as observers, leave the room. The challenge is just to understand – from our perspective as perceivers – what the primary qualities actually are. What is the basis for the secondary qualities? What remains when we remove the relational dimension, that is, that which arises through a subject's experience of the jewellery objects?

From a philosophical perspective, the distinction between primary and secondary qualities becomes salient when we look directly at the *Cosmic Debris* works. We have a visual perception of the jewellery objects. They are *there*, before us *now*. But when we turn our attention elsewhere, they are removed from our consciousness. For whom, then, do the jewellery objects exist? For whom do *meteorites* exist, before they have become localized by us here on the earth, from our egocentric point of view? We have no choice but to understand meteorites from an *endogenous* perspective – from within our system of thoughts. The meteorites, however, are *exogenous*. They come from outside our experienced reality.

The meteorite works can be interpreted as spiritual symbols, objects that cannot be completely grasped by our senses or explained fully through natural-scientific descriptive language. Ziegler has stated that he is preoccupied with myths and religious questions (Gali et al. 2013), and the *Cosmic Debris* project does raise questions about worldviews and our possibilities for gaining insight through knowledge and ritual practices. These questions are both transcendental and ontological. They are transcendental in the sense that they are about the conditions of possibility for our existence and our ability to make sense of the world. They are ontological in the sense that they concern our ability to identify the basic properties of natural-kind objects. What remains of a single exhibited object when no one's sensory apparatuses are directed towards it? Would it be justifiable to describe each individual meteorite object as constituted by stardust, or is that a relational description as well? *Stardust* is, after all, a manmade concept.

→ THE THING IN ITSELF

Given that the jewellery objects are made of meteorites, we are dually challenged. As with other objects, we can, in the first instance, make a distinction based on principle, between the relational secondary qualities we attribute to them, and the inherent primary qualities they have in themselves, despite it being difficult to describe these qualities. In the second instance, the meteorites are distanced. Because they come from outer space, they are shrouded with an extra veil of uncertainty.

The pressing problem here is the same as that famously described by the philosopher Immanuel Kant (1781), namely, that we cannot know the 'thing in itself' – *Das Ding an sich*. Kant was concerned about that which is far out there, not in a physical sense, but beyond our categories of understanding. These categories, as Heidemann (2011, p. 52) observes, can be defined as 'our forms of thinking … the pure logical forms of judgements that govern the use of concepts by the understanding'.

The forms of categorization are the mental frames we use for interpreting sense impressions, such as assumptions about time, space, causality, dimensions and substance (Gardner 1998). The assumptions are frameworks of interpretation. Kant conceives of the categories of understanding as mental glasses – they are the abstract filters we use for interpreting ourselves and the world around us. But in contrast to normal glasses, we can never take them off and see the world independently of them. Our experiences necessarily arise through a subjective perspective. The experiences are always already interpreted because the sense impressions from 'the thing in itself' must, in the first place, be categorized by our faculties of understanding. In other words, we cannot see the world in itself, uninfluenced by our thinking. As Skirbekk and Gilje (1987, p. 43) observe:

> *The unformed influences, and the thing in itself, we cannot grasp. The concept 'Ding an sich' is thus a problematic concept; on one hand, it is necessary in order to explain where experience comes from, but on the other hand, it cannot be an object for experience.*

We could try to respond to Kant's conundrum by claiming that the colours and other qualities we see are in fact part of the thing in itself (see Stroud 1995, p. 365):

> ... the qualities said to be perceived or thought about in such cases [as secondary qualities] are really qualities that do belong to objects after all. This can take the form of arguing that, e.g., the word 'coloured' just means the same as 'has the power to produce perceptions of colour in human beings'.

Could we then analogously argue that the experienced fusion crust of a meteorite is *precisely that aspect* that is the source of our experience of brown? But this would merely eschew the problem. We know something is the foundation for our experience of the jewellery objects, and that it must be separated from our experience of them. There is something that *produces* the experience of the coloured fusion crust. But what is this? What remains if we subtract our contribution to the experience?

Ziegler has given the jewellery objects forms, and we categorize them as formed. *Translucent Meteorite* (p. 22) illustrates the point. The jewellery object appears as a hybrid element – the meteorite part (Pallasite) contains both iron and the gem stone Olivin – But Kant would say this part is not a hybrid *in itself*. The concept *hybrid* is not in the object. The jewellery object has a manifestation within our categories of understanding. It appears in interpretive forms such as time and space, and we find it natural to use words like 'hybrid'. But the object itself, hybrid or not, exists independently of us. Precisely the fact that a meteorite is used as a part of the material strengthens the impression that the jewellery object will continue to exist even when no human consciousness is directed towards it.

So what were the meteorites before they entered our categorizing spheres of knowledge and understanding? In their long journey through outer space, they were experienceable, and it was, in principle, possible to form ideas about them. But *we* cannot think about the meteorites without using our concepts. How, then, can we gain access to their essence, to aspects that cannot be captured by language? The jewellery objects point to something outside the familiar world, to something beyond our horizons that we cannot, and should not, fully understand. Ziegler has said he is interested in the philosophy of Georges Bataille and his concept of transgression. For Bataille, this concept does not have an essentially negative or evil connotation; it has to do with stepping beyond the threshold of the ordinary. It is conceived of as a religious concept but not necessarily linked with any religion: we can transgress the boundaries of all forms of rational everyday activities by participating in practices that have no instrumental utility. Many of Ziegler's jewellery objects conjure associations to such practices. They can, through the mechanisms that fasten the meteorite to the rest of the jewellery object, be used in ways that remind us of ritual practices. But *Cosmic Debris* also inspires transgression in a more fundamental sense. We are challenged to think of what lies beyond our experience and knowledge. We lose control of the world we think we comprehend and master.

→ AESTHETIC QUALITIES

What about aesthetic qualities – qualities that traditionally have been central to how we categorize art? How can that which is beautiful, harmonious, or in possession of other artistic values, exist independently of our judgement of 'the beautiful', 'the harmonious' and 'art'? How can something be a jewellery object without us using the designator 'jewellery object'? Like colours, our aesthetic concepts are relational. They are meaningful predicates we ascribe to objects when we think they have aesthetic qualities.

Runa Vethal Stølen wearing
OVOID METEORITE

OVOID METEORITE – pendant – 2013 – stony meteorite NWA (Unclassified, North West Africa), silver, steel, nylon cord – 42×78×35mm

The challenge for an aesthetic realist, that is, someone who thinks there is something about an art object that really makes it an art object, is thus to explain exactly what it is about the object that causes it to have inherent artistic qualities. This problem is well-formulated by Philip Pettit (2004, p. 163):

> What has to be shown with aesthetic characterizations is their being essentially perceptual and can be explained consistently with a realistic construal … The problem, intuitively, is this. If aesthetic characterizations are held to direct us towards real properties of the works they characterize, how then do we account for the rather unusual nature of those properties?

The explanatory challenge is the same as with colours. If aesthetic qualities have a realistic existence – if they are properties some objects have in themselves – then they must apparently be primary. But as Pettit points out in the quote above, it seems like aesthetic qualities are relational and therefore secondary. They appear as human forms of evaluation. So how should we explain the fact that we have a fundamental intuition that art objects possess *intrinsic* aesthetic qualities?

This question transcends traditional disputes about how specific art concepts should be applied. A debated question is which aesthetic concepts are right to use about jewellery objects and art jewellery, and Ziegler has himself participated in this debate (Gali et al. 2013). But independent of the question of how one should describe Ziegler's works and other forms of art jewellery, there is a more general point to be made: to be a jewellery object does not seem to be an *essential property* an object can have. Like colours and other secondary qualities, to be a *jewellery object* is a relational quality – it is something an object has by virtue of us calling it 'jewellery object'.

The problem is brought to our attention in *Cosmic Debris*. Some of the meteorite works are made in shapes often associated with traditional jewellery. In *Ovoid Meteorite* (p. 29), the meteorite has been given an oval form that creates a deep space for interpretation. On one hand, the oval did not come to us from outer space. It is carefully shaped by human hands and placed within a human aesthetic context – a categorized reality – that is emphasized by the ball-joint fastening mechanism the wearer controls. On the other hand, the jewellery object appears pure and expanding. The oval creates associations to something genuine and original, to archetypes and life-giving forms in a cosmological space. As with several of the other jewellery objects, the base material is chondrite – a stone meteorite composed of chondrules, the very building blocks of our solar system. The chondrule grains in the object are highlighted and point back to a time when everything was different, not least from our perspective.

The same is the case for *Meteorite Ball* (p. 10). Its spherical shape draws associations to a seed, to an origin, and the word chondrule in fact means seed or small grain in Greek. But 'seed' is a word we have made up and it holds our associations. The chondrules are there but are not thought of as *seeds*. So tension arises in the interpretive perspective. The associations to seeds and the origin of meteorites are based on our natural-scientific knowledge about the world. This knowledge pulls the jewellery objects towards us. But in the next instance, that same knowledge pushes them away. We are humbled by the distance in time and space to something so much larger than ourselves.

Again, questions arise: What is fundamental in the universe we live in? What is left when we peel away our categorizations? What is the transcendental purpose of this life we have been given, and the role each and every one of us has here on this little planet Earth? The questions open a space for ideas that burst the boundaries of natural-scientific and reductionistic explanations.

→ BEYOND CONCEPTS

The distinction between primary and secondary qualities can be analysed from different perspectives. One common way of elucidating the distinction is to do it conceptually – by using concepts from modern philosophy and metaphysics. In the field of philosophical art theory, the goal of analysing the concept of art has often been to clarify what art is in general, as a concept and ontological category, independent of the diversity of personal opinions and subjective evaluations that vary from person to person (Nordby 2005).

It is however also possible to analyse basic aesthetic qualities non-conceptually, in a more subtle, nonverbal way. This strategy would involve *showing* something to be the case, rather than explaining facts in the external world in terms of words that literally refer to them. Ludwig Wittgenstein, probably the most influential language philosopher of the last century, is famous for having marked a distinction between making qualities apparent with the help of descriptions, and making them apparent by non-linguistic, aesthetic means:

There are, indeed, things that cannot be put into words. They make themselves manifest (Wittgenstein 1922, p. 73). What can be shown, cannot be said (ibid., p. 26).

This does not mean Wittgenstein thinks aesthetic qualities have an *objective*, universal existence that can be shown. In his later philosophy, Wittgenstein (1966) is particularly keen to stress that achieving aesthetic insight is not a matter of uncovering general facts. Instead, he argues that aesthetic knowledge is culturally conditioned, rooted in multiple ideological assumptions:

The words we call expressions of aesthetic judgment play a very complicated role, but a very definite role, in what we call a culture of a period. To describe their use or to describe what you mean by a cultured taste, you have to describe a culture. What we now call a cultured taste perhaps didn't exist in the Middle Ages. An entirely different game is played in different ages (Wittgenstein 1966, II, p. 8).

Wittgenstein underscores that things conceived to be aesthetically valuable within *one* culture – within what he calls a family of language games – cannot be linguistically described as universal truths. To understand what it is that has aesthetic value, we must, he says, show it through observable action. As Snævarr (2008, p. 155) observes, aesthetics, for Wittgenstein, is linked to the carrying out of practices:

According to this school [the Wittgensteinian tradition], the core concept of aesthetics is aesthetic practice. We 'traffic' in artworks, acting on their behalf in diverse ways. The world of art consists of loosely linked practices which are, in turn, linked with non-aesthetic practices. One of these practices can be to paint, another is to interpret and evaluate paintings.

Seen in this light, the practices of making and exhibiting art jewellery also constitute a field for aesthetic evaluation, as long as they are practices that deem jewellery to be art. That which is aesthetic must be understood as such by being based on concrete patterns of action; it cannot be understood from the opposite approach, in light of unified facts in the world. In contrast to what many have thought about the reality of art, Wittgenstein claims there is no single general criterion for determining whether something is art and which can include the entire set of aesthetic practices.

This point – about the absence of a general criterion for art – generates a paradox for Wittgenstein. If the question – What are the right and wrong uses of aesthetic concepts? – must be related to divergent practices, how can it be objectively valid to maintain that knowledge and aesthetic insight are relative concepts? The claim 'aesthetic knowledge is relative' is put forward in a philosophical language game, so would not the opposite claim 'aesthetic knowledge is not relative', when put forward in another language game, be just as valid? This is the famous self-referencing paradox in philosophy: if you claim that knowledge and truth are relative concepts, how can the claim 'knowledge and truth are relative concepts' itself be universally valid (that is, non-relativistic)?

Early on, Wittgenstein (1922) claims that in the final analysis, the only meaningful sentences are those that refer to concrete states of affairs in the world. Since his own theoretical arguments in support of this claim are abstract and therefore do not refer to concrete states of affairs, he comes ultimately to the conclusion that his own arguments are meaningless. But how could they then be meaningful and communicative as philosophical arguments? Wittgenstein's (1922, p. 74) solution to the paradox is the famous ladder metaphor: when we climb up the ladder that provides us with general knowledge, when we have gained the abstract insights, we must throw away the ladder:

> *My propositions are elucidations in the following way: anyone who understands me eventually recognizes them as nonsensical, when he has used them – as steps – to climb up beyond them. He must, so to speak, throw away the ladder, after he has climbed up it. He must transcend these propositions, and then he will see the world aright. [He will then understand] … that what we cannot speak about we must pass over in silence.*

After studying *Cosmic Debris* as well as Ziegler's earlier works, I have the strong impression that his project falls under Wittgenstein's aim of giving a *silent presentation*. The meteorite works can, in an aesthetic way, help us shed light on the question of what aesthetic qualities really are. But the answer to this question does not come to us through explanatory concepts. It comes instead to light through aesthetic forms of expression that cannot be reduced to communicative noise – verbal statements or other forms of normal descriptive language.

One of the works can, in fact, be understood as commenting on this strategy. In *Meteorite Chisel* (p. 34), Ziegler has given the meteorite a chisel shape. He has been inspired by an iron meteorite at the Natural History Museum in London, which is shaped like a knife. It was actually shaped to function as a knife by Inuits in north-west Greenland in the early 1800s, presumably because they realized how strong the material was, yet without being aware that it was a meteorite. The difference between the meteorite knife and Ziegler's chisel is that the knife, functionally speaking, really is a knife. Ziegler's intention, by contrast, is to use the tool metaphor as an artistic means. For him, the jewellery object gains a self-referring function analogous to Wittgenstein's conceptual ladder metaphor. The work itself comments on the craft and skill with which it was made. When a meteorite is used for this purpose, and when the object is a jewellery object, this happens by virtue of an aesthetic logic that cannot be reduced to explanations or other linguistic disturbance.

→ PRIMARY AESTHETICS?

Yet the point of *Cosmic Debris* goes deeper. Ziegler is not just interested in using aesthetics as a strategy for showing us something. He is also challenging the idea that aesthetic qualities must be secondary. As I have already argued, aesthetic *concepts* are necessarily relational. They must be so since they are part of our horizon. It is we who create the explications of the concepts. Based on this realization, most philosophers of art have concluded that the aesthetic qualities of art objects must therefore also be relational. But we must not let the distinction collapse between the level on which we *think* about the world, and the world in itself. It is not given *a priori* – independent of our sense experience – that all aesthetic qualities are relational.

I interpret *Cosmic Debris* as a fundamental attempt to show that at least some aesthetic qualities are primary and exist beyond our categorized experience. In other words, the artist's goal is not simply to present us with secondary qualities. His more ambitious goal is to accentuate aesthetic primary qualities – qualities linked to aesthetics that burst the boundaries of what we can describe. It is not possible to state what these qualities are because then they would be relational, captured within our conceptual apparatus. But indirectly, we can say that Ziegler wants to show us *pictures* of something beautiful and genuine, which lie beyond our ability to use language-based definitions.

Again, it is appropriate to draw a parallel to something Wittgenstein says in his early philosophy. He claims that thoughts about facts represent the world inasmuch as they are pictures that stand in a parallel relation to states of affairs – arrangements of objects:

The logical picture of the facts is the thought (Wittgenstein 1922, p. 10).
In a picture the elements of the picture are the representations of objects (ibid., p. 8).

According to Wittgenstein, it never is possible to describe thought pictures such that the descriptions become as rich as reality. The decisive nuances will always be missing. In the same way, Ziegler tries to get us to turn our gaze to something farther out, larger and more extensive than ourselves. Analogous to how we are helped to understand something by looking directly at a picture, so also are we helped by *looking* directly at the jewellery objects.

→ USING AESTHETIC CONCEPTS TO APPROACH ART

What does it actually mean to use concepts to analyse Ziegler's art and find it meaningful? This can be a strategy for

METEORITE CHISEL – pendant – 2013 – stony meteorite NWA (Unclassified, North West Africa), silver, nylon cord – 130×20×15 mm

moving closer to the primary qualities linked to nature and nature-related practices. Yet there are two crucial differences between this type of conceptual approach and a more fundamental aesthetic analysis of primary qualities. First, rituals and practices are activities involving people. Primary qualities, by contrast, cannot be actions. They are not practices we can describe. If existing at all, they exist independently of human activities.

Second, concepts are reductionistic in a psychological sense. They constitute mental cubicles or 'bookshelves'. Their categorizing function pulls the jewellery objects towards us, away from their primary existence and aesthetic uniqueness. Concepts, as Burge (1979) observes, are 'our means of mental representation. They are ways of thinking of objects and properties'.

As long as the jewellery objects in *Cosmic Debris* are not modes of thought in themselves, our conceptual understanding of them will never be complete. The objects can never be identical to our concepts. Nevertheless, we have a need to categorize our attempts to represent the objects. The objects in thought can be linked to communication and the goal of explaining to ourselves and others what something is. Human communication is an activity, the goal of which is to mediate something, and the activity entails using language, which involves concepts. But when the goal is just to understand, language is not necessarily crucial. Perceptual realists would say we can see and understand the empirical world directly, and that we have direct access to that which we encounter through our senses. We can 'link up' with the jewellery objects through perception. Ziegler emphasizes this on-and-off linkage with the fastening mechanisms of several works. They give us as wearers the possibility of moving the meteorites this way and that, as a self-defined practice. Still, the possibilities for movement are limited by the particular fasteners (carabiners, cardan u-joints, ball-joints, etc.).

In a far more fundamental way, however, the movement limitation, the space for action, is intimately linked to *our horizon* of understanding, to the overall perspectives we have of ourselves and the surrounding world. This horizon is the total 'amount of beliefs and opinions we have at a given point in time, whether we are consciously aware of them or not' (Føllesdal and Walløe 2003, p. 31). It includes everything that we use to understand ourselves, others and the world around us. As such, a horizon of understanding is necessarily unique. Two people's horizons will never be completely identical, since no two people have precisely the same thoughts and opinions.

As long as meteorites are entities we have limited objective knowledge of, we as wearers of the jewellery will form different subjective interpretations of them and their significance in a more comprehensive perspective. When Ziegler so strongly stresses the linkage to a particular person – the wearer – he accentuates how person-dependent the works are. So the subjective connection is twofold and profound: the wearer is both physically and cognitively attached to the meteorite work. Moreover, this subjective dimension highlights the fact that there is little we humans actually have of *common* knowledge. The meteorite parts of these jewellery objects appear as things we, at least ideally, can try to understand in the same generally valid way, based on a shared language, yet with the knowledge that shared understanding will always only remain an ideal. As the limitations of our mental and analytical resources become salient, we as wearers withdraw from the limelight.

→ A THIRD-PERSON PERSPECTIVE

The point is quite general: having different interpretative horizons hinders us from having jointly-shared language-based insight. Two people will never understand a linguistic expression in exactly the same way. The question is therefore to what extent two people must understand an expression in order to associate it with the same concept, in order to exchange ideas and opinions with the same content (Nordby 2004). When the expressions 'art' and 'art jewellery' are used in statements as descriptive nouns, a double challenge arises. First, it is problematic to reduce the aesthetic object to language ('This *is* art'). Second, the statement can function as a fixed category that impedes a genuine encounter, one where an artwork truly appears as an artwork in all its open presence.

If, however, we are intent on identifying what we have defined as primary qualities, we should look for something more fundamental. The goal is to reveal something that exists independent of our respective perspective-based understandings, something that is separate from personal, idiosyncratic descriptions. The American philosopher Thomas Nagel (1986, p. 209) calls this the aim of adopting a *third-person perspective* – it would enable us to see ourselves from outside, as others see us. The challenge is just that this, for each of us, is 'a view from nowhere':

> ... the detached view of our own existence, once achieved, is not easily made part of the standpoint from which life is lived. Far from enough outside my birth seems accidental, my life pointless, and my death insignificant, but from inside my never having been born seems nearly unimaginable, my life monstrously important, and my death catastrophic.

We are our own world. We see the world as if through the base of a funnel. It widens out for us, not just in our field of vision, but also more generally, through the thoughts and beliefs we have about our environment. In *Meteorite Cone* (p. 15), Ziegler highlights the concept of expansion by giving the meteorite a conical shape. We as wearers can view it from below, but also from the opposite standpoint: from above. Conceptually, this compares with trying to adopt a third-person perspective. Viewing the cone from the outside corresponds with trying to imagine how each of us is reduced to one point amongst many others in the universe. In this case, we as wearers lack a conscious connection to the standpoint we take. The only possibility is to *indirectly* represent a third-person perspective in our own consciousness. None of us can climb out of ourselves and see ourselves from the outside – we always see things from within our own respective horizon. And even those who can see us from the outside only have indirect access to our private consciousness, to our thoughts and beliefs.

But the tension between the first- and third-person perspectives does not simply relate to the possibility of our knowing ourselves and the external world. It also has an existential aspect. As Nagel (ibid., p. 205) observes: 'when we try to see ourselves from the outside, we find it hard to take our lives seriously. This loss of conviction, and the attempt to regain it, is the problem of the meaning of life.'

From the outside, each of us is just one of five billion equally important individuals, more or less randomly placed on a planet that has objects and events that, in a cosmological perspective, are at least as significant as we ourselves are. We neither can nor *want* to adopt a third-person perspective of ourselves and the world around us. That would be the same as managing to climb outside of our-

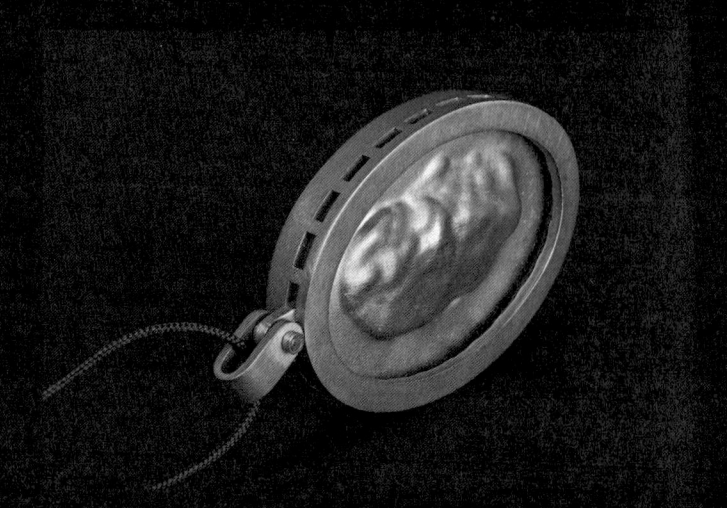

EMBEDDED METEORITE – pendant – 2013 – stony meteorite Chergach (El Mokhtar, Mali), silver, nylon cord – 78 × 47 × 20 mm

selves. But Nagel's point is that to adopt a third-person perspective can nevertheless be an ideal, an attempt to understand something from an objective and *just* standpoint. It can also be an attempt to understand what it is that has value for us from a more objective standpoint than we can have on our own, individually. That which has value from this super-ordinate perspective must be *shown*, since all language-based descriptions must be made from a first-person perspective. Aesthetic forms of representation can therefore be meaningful because they are not dependent on linguistic description. I understand the goal of communicating values from outside ourselves, in an aesthetic way, as a basic intention of *Cosmic Debris*.

→ THEORY OF CAUSATION

The meteorites we find here on earth have hit us. They have entered the first-person perspective. Ziegler emphasizes this by stressing the speed the meteorites achieve by the time they hit the earth's crust. In the pendant *Embedded Meteorite* (p. 39), the force of impact is marked by the impression of the meteorite on the other side of the plate to which it is fastened. The meteorite has not just entered the earth's atmosphere; it has gone directly to the wearer and impacted that individual. Yet it has stopped on the brink of the human body's boundaries. It has almost broken through, but the intervening material has held it at bay. The plate between the meteorite and the wearer's body can be understood as a link between two conceptual antithetical poles. The meteorite's primary qualities, in terms of knowledge, are so far removed from our horizon of understanding as it is possible to get.

In the first instance, the meteorite has its special qualities like all other objects in the external world. But as a meteorite, it is swathed in an extra layer of mystery, and even what we think we understand seems distanced and detached from our ordinary life world (*Lebenswelt*). In *Translucent Meteorite* (p. 22), the small transparent olivine crystals are highlighted as a reminder of a chemical process that occurred long ago and far away from where we are. We think we understand the process but it is obscure, and our knowledge is uncertain.

Nevertheless, we are the centre of the world from our own standpoint, with a richly-endowed consciousness and egocentric perspectives. We wearers have privileged first-person access to our own consciousness of ourselves. We are 'bombarded' by our own thoughts and beliefs, for good or ill. None of us can eschew our own thoughts. Even when we actively try not to think about something, we often cannot avoid doing so. (When I was a student, a teacher once said: 'Try not to think of pink elephants'. It was absolutely impossible!) Even though the meteorites have come to the body, they have not entered the *inner* realm of thought. In addition to the great knowledge-related distance to the primary qualities, there is a physical barrier between our minds and the meteorites.

Meteorites in the Formation of Cassiopeia Constellation (p. 43) further examines the relation between our horizon of understanding and the external world. It contains five meteorites set in an asymmetrical structure that falls under no simple form-category such as an oval or ball. But as the title divulges, it has a form that is recognizable to us as the constellation *Cassiopeia*. The constellation in itself does not exist as a real unit in outer space. The stars are individual objects we have linked together, not because we need to give them a reductionist explanation, but in order to orient ourselves in a larger context.

The need to gain a sense of coherence does not only arise when we create connections between objects in the external world. The tension also arises internally in our horizons. Many of our beliefs are connected through mental networks that are not necessarily logical. The perspectives we have of ourselves and the surrounding world are seldom fully rational. They often contain contradictions – beliefs that are inconsistent with each other. As we strive to conform to norms of rationality, we often try to understand how inconsistencies can be avoided, how we can create meaningful mental schemes.

Similarly, all the meteorites in *Meteorites in the Formation of Cassiopeia Constellation* are unified entities that more or less randomly have come to us, in the same way as specific thoughts suddenly emerge in our consciousness as defined mental episodes. And just as we have not decided that the meteorites should come to us, neither can we de-select the thoughts that come to us in the first-person perspective. We can have a critical relation to our thoughts, but the thought contents are there, regardless of whether we want them or not. And viewed in isolation, a single thought episode is never inconsistent and incommensurable with itself. It is just there, present to the mind as a conscious episode.

This point also extends to objects in the external world: viewing a star or meteorite in isolation from its place in a larger system, it does not make sense to say that it can fail to fit into that arrangement. It is just there, where we see it. Yet there is still a crucial difference between thoughts and external-world objects: thoughts can make a negative impact *in* our horizon. Particularly negative thoughts can seem like ruthless, brutal mental experiences. It is often difficult to avoid such states of consciousness. And even when we manage to supress them for a time, they suddenly recur, against our will, stronger and more intense than ever before.

That said, negative thoughts can disappear once we see connections into which they may fit. Analogous to how we create connections between objects in the world – such as when Ziegler makes a star constellation – we create connections between things in our own horizon of understanding. Ziegler reminds us that both activities are attempts to master the world. To make a star constellation is an attempt to place ourselves in a larger context – we draw the space of the world closer to our recognizable concepts and need to understand. To create connections in our inner perceptions is an attempt to avoid dissonance – tensions between divergent beliefs (Nolen-Hoeksema et al. 2009).[3]

This has a fundamental ideological dimension. We humans are used to understanding and mastering most things. It is difficult for us to accept that there are thought processes and areas of knowledge we cannot control. Likewise, it is difficult to exercise linguistic power over the jewellery objects in *Cosmic Debris*. We are unable to control them by capturing them inside our system of concepts. Some of the jewellery objects, for instance *Widmanstätten Structured Meteorite* (pp. 19, 20) and *Weathered Fusion Crusted Meteorite* (p. 9), have natural, un-manipulated shapes, and this underscores the point: we experience huge problems trying to use our categories of understanding. We who seek knowledge, and anyone who wears the jewellery objects, recede even further into the background.

METEORITES IN THE FORMATION OF CASSIOPEIA CONSTELLATION
brooch – 2013 – stony meteorites NWA 869 (North West Africa), silver, steel – 120×53×27 mm

→ EFFICACIOUS FORCES

The effort to understand this metaphysical dimension of the jewellery objects falls under a modern direction in the philosophy of mind that is concerned with levels of causal explanation. These are theories about how it is appropriate to assign causal power to objects – about how they can be part of causal chains – all in relation to on our explanatory needs, interests and goals. The basic idea is that it is correct to assign an object a causal property if doing so coheres theoretically with what we otherwise believe. The idea about coherence, as Føllesdal and Walløe (2003) observe, is that the assigning is true if and only if it fits into a coherent and all-encompassing set of beliefs.

The starting point for philosophical analyses has often been a modern version of Kant's aforementioned idea that everything we see is formed by our categories of understanding (e.g., the manifold of time, space, nearness and distance). Hence, it seems difficult to ignore our contribution to the world – those things we subjectively add to pure sense impressions – and to cognize the actual causes of our experience. Many philosophers have therefore assumed that factual causes, the qualities that really cause something, cannot be relational (Crane 1991). When one object causes another object to move, there must be something about the object itself that constitutes the cause, not the object's relation to a viewer. David Hume's billiard balls are a famous example (Stroud 1977). We see one ball approach another, we sense that we see its impact, and we see the second ball begin to move. But all this is relational, and as Hume points out, we do not see anything additional that would appear as the cause in itself. So how should we understand what the activating force is? The question surfaces regardless of how we try to delimit the cause as being a quality that 'underlies' something, as the causal starting point for our sense experience.

As Jackson and Pettit (1995, p. 229) observe, the question is completely general. It pertains to all qualities that can have causal force. Jackson and Pettit use the example of explanations of human action and the idea that it is the content of our beliefs and desires that is the precipitating cause of the actions we perform. This content, however, is relational – it depends on the role our beliefs and desires have. How, then, can mental content cause anything at all, if only non-relational qualities can be causes?

What one neurophysiological state does to another neurophysiological state is a function of their relatively intrinsic character, not of their relations to more remote states, regardless of whether those more remote states are inside or outside the skin. How then can the possession of a certain content by an agent's intentional state – a highly relational matter according to functionalism – explain the agent's behavior?

The conundrum applies to all causal chains we think can be objects for visual experience. When we think we 'see' a cause, we see a movement. But if causal forces are non-relational, then we cannot see the inherent cause, as long as all we see, *per definition*, is relational.

The point comes to expression in the fastening mechanisms and pendants in *Cosmic Debris*. They are mechanisms for change – causal forces in a system. We can use the mechanisms in repetitive movements, as objects of intentional actions that are not linked to normal, everyday activities. When the movements are transferred to the meteorite parts, we gain an increased sense of transcending ordinary practices. But as far as sense experience is concerned, we still lack access to the efficient cause for the movement. For example, the top of *Ovoid Meteorite* (p. 29) consists of a rod attached to a sphere (a ball-joint). It seems apt to say that when we press the mechanism, we make the angle of the meteorite change. Similarly, when we manipulate the u-joint fastener on *Meteorite Cone* (p. 15), we think we 'see' the conical shape of the meteorite change appearance in time and space. But what we describe is only relational. What really happens, knowledge-wise, is as far removed from us as the meteorite's origin.

A striking aspect of Ziegler's project is that the distance between our interpretational frames and the jewellery objects' primary causal qualities creates epistemological tension: interpretations and our use of the objects create personal nearness, but the jewellery objects point beyond themselves to exogenous causal processes and events – so much so that we find them overwhelmingly difficult to grasp. And it is not even our mental powers that have brought the meteorites within our horizon. It was gravitational force that caused the meteorites to come to earth.

→ THE EXTERNAL WORLD

Kant argued that the qualities we think we can name through concepts are relational, since concepts are categories of understanding we place on objects. Thus, when we call something a brooch, it is our conceptual category *brooch* that we use. The thing is a brooch for us but not in itself.

Many philosophers have maintained that Kant's theory raises a deep philosophical problem. For if Kant is correct, how can the objects we think we refer to really have the qualities we ascribe to them? When we ascribe a concept such as *jewellery* to an object, then we mean what we say – that the object really has the quality of being a jewellery object. And when we say that *Meteorite Ball* has a spherical shape, we mean it really has a spherical shape. But how can attributions of concepts such as *jewellery object* and *spherical shape* ever be valid? If they are relational concepts, are we correct in claiming that these objects really have the qualities we are referring to?

It would be counter-intuitive to accept that we *always* are wrong when using concepts to ascribe qualities to objects. The solution to this problem is to start at the other end. We can instead say that as long as we are not always massively mistaken, the relation between concept application and the qualities must be valid, both when we use aesthetic and non-aesthetic relational concepts. Viewing *Cosmic Debris*: it must, in some sense or another, be correct to say that the Widmanstätten pattern truly is beautiful, and that the fastening mechanism of *Ovoid Meteorite* (p. 29) does in fact cause the angle to change when we push and turn it. Nor can it be incorrect to claim that the meteorite truly has an oval shape. The problem is just to understand how these claims can be true.

We can avoid the problem if we relate the external world's existence to our ordinary explanatory practices. As long as we, in our practices, accept that the world has an existence independent of our categories, then that is the case: that which we regard as true and real cannot be determined just by using a general and abstract philosophical theory, but by what we normally regard as true and real in various language-based discourses. It is these discourses that determine how reality is, not ontological states of affairs. Instead of starting out from metaphysical assumptions, we should start from the other end, from what we believe. In so doing, we take for granted our ordinary practices of distributing secondary qualities.

Again, this point stems from Wittgenstein's late philosophy (1957). Wittgenstein argues that whatever is true and false must be related to what he calls language games – to the different ways we use language systematically to talk about ourselves and the world. We can use the challenge of explaining intentional action, as presented by Pettit and Jackson above, to illustrate: if we want to explain a person's actions in a psychological vocabulary, we give a psychological explanation. We explain the person's behaviour by saying he had such and such beliefs, that he had such and such desires to achieve something, and that this is what lay behind the action he performed. Hence, when the wearer of *Meteorite Ball* (p. 10) makes the ball turn, we would normally explain his behaviour by attributing to him the belief that it was possible to turn the ball, and that he wanted to make it turn. As long as the explanation works – as long as we achieve the goals set for the explanation – then that particular belief and desire exist as genuine causes in the world.

Similarly, if we, in a different practice – a different language game – want to understand the physical mechanisms underlying the wearer's behaviour, we would give a physiological explanation: it was such and such physical chains of cause and effect in the wearer's body that resulted in the behaviour. When the wearer of *Meteorite Ball* spins the ball, there are physical causal forces underlying the fact that the ball spins. The spinning is the final effect while the goal-oriented action starts at the other end, as neurophysiological processes in the wearer's brain. Again, as long as the explanation functions, then the explanation is correct.

The point is simply this: what we regard as genuine reality depends on our interpretative framework. And within our normal interpretative frames, the world exists, as it is and as it has been, independently of ourselves. We usually think dinosaurs existed even though no people were around to call them 'dinosaurs'. We assume that mountains, seas, continents and outer space exist around us, even though we ourselves have defined the concepts we use to talk about them. The same is the case for the physical material from which Ziegler's jewellery objects are made. This material is *there*. And what we see appears as real *for us*, within our interpretative space. Thus the burden of proof lies with the sceptic who thinks that what we see is not real. We have an experience of Ziegler presenting the material as something exemplary, with aesthetic qualities. And when the aesthetic qualities stand out as real, then of course the more foundational physical qualities become real. To emphasize primary aesthetic qualities in the way Ziegler does constitutes a basic challenge to anti-realists who hold that we cannot access the external world through language, perception and action.

→ NORMATIVE ASPECTS

Is there any ideological message in Ziegler's art? Could it in any way suggest to us how we should act? It would be an exaggeration to say that *Cosmic Debris* has moralizing content, but the meteorite works do have a clear ideological dimension: the tension between the interpreter's perspective, the references to transgression and the exogenous third-person perspective translate into an ethical dilemma between what Nagel (1980) describes as the personally near and morality's distanced demands. As an illustration of the dilemma between nearness and distance, Nagel gives (ibid., p. 190) an everyday example most Norwegians can relate to personally:

> *To take an example close to home: the bill for two in a moderately expensive New York restaurant equals the annual per capita income in Bangladesh. Every time I eat out, not because I have to but just because I feel like it, the money could do noticeably more good if contributed to famine relief. The same could be said of many purchases of clothing, wine, theater tickets, vacations, gifts, books, records, furniture, stemware, etc.*

Nagel understands this as an ethical dilemma about choosing between something experienced as subjectively good, and morality's stricter demand to take into consideration a just distribution of goods for the earth's population as a whole. Nagel's own position is that the subjective perspective has a certain validity, but that it must be balanced by holistic, universal principles (ibid., p. 209).

> *… doing the right thing … is not the whole of it, not even the dominant part: because the impersonal standpoint that acknowledges the claims of morality is only one aspect of a normal individual among others … there is much more to us, and therefore much more to what is good and bad for us, than what is directly involved in morality.*

The crucial distinction here is between *experiencing* that you are doing good, and that of meeting objective requirements for good actions. While classic moral philosophical theories have not accepted this distinction – to do something good has been understood as to follow morality's demands, which emanate from somewhere external to the actor – Nagel argues that we need both 'the inner and outer perspective' (ibid., p. 197). The demands to do something subjectively good are not as strong as the external moral demands, but they are just as legitimate – as long as they are not too dominating. We should not, therefore, have a bad conscience simply because we do not fully live up to objective standards.

There are similarities between Nagel's analyses of morality's demands and Ziegler's focus on that which lies beyond us and which is originary. Ziegler is also preoccupied with a third-person perspective, and he also has a normative message: in order to find an *existential* balance, and in addition to using the first-person perspective, we must try to see ourselves from the outside. At the same time, there is a decisive difference between Nagel's philosophy and Ziegler's artistic intentions. While Nagel focuses on people and ethical ideas about justice and equality, Ziegler applies a more comprehensive cosmological and ethical perspective. Nagel, moreover, believes we can describe right and wrong from a third-person perspective, that we can formulate the objective moral demands as clear *arguments*. For Ziegler, the goal is also to direct attention to something beyond our endogenous perspectives, but this is not something we can describe fully or codify with concepts. The aesthetic contents in *Cosmic Debris* cannot be reduced to arguments and analyses.

There is nevertheless something normative and value-laden about the project. We are reminded that we should not be as egocentric as we tend to be. The point is underscored in *Square Meteorite in a Circle* (p. 49). The jewellery object is shaped like a classic mandala – like a square inside a circle. The mandala traditionally represents wholeness and unity; it represents every living thing in a comprehensive, cosmological perspective. A mandala is often used as a symbol of the human self and life itself. It is a cosmic diagram and a reminder of our relation to that which is eternal, to a world that stretches far beyond ourselves. The famous psychoanalyst Carl Gustav Jung used the mandala concept for what he thought were eternal archetypes arising through dreams. He saw them as unconscious processes. Ziegler has earlier stated that he is interested in Jung and other thinkers who interpret fundamental human experiences as expressions of existential insights about which we have limited conscious knowledge.

Independent of the various ways in which it is possible to interpret the mandala symbol conceptually, Ziegler's use of it has a more fundamental dimension: when the symbol is used in a jewellery object, it gets close to, and draws attention to, a particular wearer. The personal nearness and the distance are linked; the individual encounters the all-encompassing cosmological ideas underscored by the meteorite material. But neither the wearer nor other parts of the world become more important than any other elements in the cosmological sphere. To say that Ziegler is a philosophical pantheist – that he is committed to the idea that nature is besouled – would be to go too far. But in an eco-philosophical perspective, *Cosmic Debris* clearly fits into a theoretical direction emphasizing ideas of sustainable development and stewardship of the world in its entirety. The world gets out of balance if the first-person perspective becomes too dominant.

Ziegler's project ultimately can be interpreted as a confrontation with *individualism* as an ideology. In our modern era, individualism and ethical egoism hold strong positions (Nordby 2013). We are used to thinking about and prioritizing ourselves and that which is near to us. We feel it is acceptable to be morally myopic and to favour personal gain rather than justice, solidarity and equality (ibid.). *Cosmic Debris*, first of all, is a confrontation with individualized art jewellery. Ziegler's art is fundamentally in opposition to art jewellery as decoration and as symbolizing the wearer's singularity. But his project is ultimately also a confrontation with individualism as a more general ideology permeating all human action.

→   LANGUAGE GAMES

The jewellery objects, however, do more than express normative ideas. They are themselves expressions and, in a wide sense, can be interpreted as language, as symbols with intentional content. The crucial difference between normal words and Ziegler's jewellery objects is that the jewellery objects are not conceptual and literally self-referential. When they highlight aesthetic qualities, they can be understood as aesthetic *pictorial expressions*.

Understood as visual representations, they refer to a transgressive reality that is not ours. Some philosophers have held that since there is a reality to which we lack access, there must be a divine being. This is one of Descartes' famous arguments for the existence of God (Rokka 2008). Now it is probably not Descartes' philosophy that has been the basis for pointing out the religious undertones of Ziegler's art, but *Cosmic Debris* clearly invites contemplation on spiritual and transcendental questions about the preconditions for our existence. And since the jewellery objects are aesthetic objects, it seems correct to address these questions outside traditional natural-scientific frameworks.

SQUARE METEORITE IN A CIRCLE

pendant – 2013 – iron meteorite Muonionalusta (Norbotten, Sweden), silver, ebony, nylon cord – 90×6 mm – private collection

Given how Ziegler's jewellery objects can be interpreted as pictorial representations, they can also be understood as posing a challenge to some contemporary art jewellery. Ziegler has previously stated that he does not identify with conceptual-critical traditions that make 'jewellery about jewellery' (Gali et al. 2013). But he also marks distance from what he has called modern conceptualism, even though, in one sense, he sees himself as a conceptualist along the lines described above. However, here the qualifier 'in one sense' is important. The goal is not so much to express specific ideas as to address the question of what it means for ideas and concepts to exist at all.

As I experience the intentions of *Cosmic Debris*, its works support that which modern language philosophy calls conceptual Platonism. This philosophical current sees concepts as existing 'out there', as Platonic ideas; they are things we hunt and gather – we gradually understand them better, as we develop an implicit understanding of their content (Peacocke 2008). Conceptual Platonism is often linked to the founder of modern language philosophy, Gottlob Frege, who held that we never can know if we have a thorough understanding of the language we use (Wright 1991). We steadily approach a correct understanding, but we never know if we have understood our concepts correctly. Concepts are anti-individualistic entities 'outside our heads'; they are more or less unknown entities, just as are classic Platonic ideas (Taylor 2001).

For me, the central themes of *Cosmic Debris* can be understood in light of this thinking. The jewellery objects are transgressive; they remind us that there is much we do not know, and that we should be humble about the knowledge we think we have. But if we have a humble relation to the world, we should also have a humble relation to what our language means. It is what our language refers to in the world that determines the content of the concepts we possess and use when we think and express ourselves. As such, it is not we who have power over language. The concepts our language expresses exist as external ideas we steadily uncover, if we are on the right track. But we can never know for sure that we have a complete understanding of true reference, not even that we really are on the right track.

This is a general theory about concepts and concept possession, and it will thus also extend to aesthetic concepts and the concept of art. Many thinkers have tried to define art, and most of the attempts have been reductionistic. The goal has been to develop a humanly-created definition for what it is that falls under the concept 'art' – to bring the concept down to earth as a fully understandable entity (Nordby 2005). Ziegler's works can remind us that this might be impossible, that art and the concept of art are essentially anti-reductionistic, anti-individualistic, and, in a metaphorical sense, like meteorites: the concepts come to us, but they have aspects that transcend our capacities for knowledge. If this is a plausible hypothesis, the tables are turned: it is not we who decide what art is, but the concept of art that sets conditions for us.

NOTES

1. I would like to thank Reinhold Ziegler and Arlyne Moi for their helpful comments to earlier versions of this essay.
2. Ziegler's artistic practice has previously been discussed with the interpretive frames of art history and religious currents such as shamanism and pantheism. His works have been linked to ritualistic actions, myths and primitivism. These are indeed apt approaches, but since others have covered this ground, I choose instead to approach Ziegler's *Cosmic Debris* series from philosophical perspectives and simply refer the reader to Gali et al. 2013.
3. In itself, the experience of great tension can cause us to change our perception, even without rational grounds. One example can clarify this idea: if, on one hand, we perceive that the Norwegian winter is terribly long, cold and dark, but on the other hand, we perceive that it is here we will remain living, we will try to do something about the tension. Most of us cannot do anything about the fact that we live in Norway. To get around the unbearable dissonance between how we feel and how we would actually like Norwegian winter to be – day after day throughout the whole winter – we have a defence mechanism that causes us to change what we actually can change, namely, our perception of winter in Norway. Without actually having a good reason for changing our opinion, we suddenly feel that winter in Norway is ok.

BIBLIOGRAPHY

Burge T. (1979). Individualism and the Mental. *Midwest Studies in Philosophy*, 4.
Crane T. (1991). All the Difference in the World. *Philosophical Quarterly*, 41.
Smith, C., Russell, S. & Benedix, G. (2009). *Meteorites*. London: Natural History Museum.
Føllesdal, D. & Walløe, L. (2003). *Argumentasjonsteori, språk og vitenskapsfilosofi. [Argumentation Theory, Language and Philosophy of Science]* Oslo: Universitetsforlaget.
Gali A., Hölscher, P. & Henriksen, H. (eds.) (2013). *Aftermath of Art Jewellery*. Stuttgart: Arnoldsche.
Gardner, S. (1998). Kant. In A.C. Grayling (ed.), *Philosophy: A Guide through the Subject*. Oxford: Oxford University Press.
Heidemann, D. (2011). Understanding: Judgements, Categories, Schemata, Principles. In W. Dudley (ed.), *Immanuel Kant: Key Concepts*. Durham: Acumen.
Kant, I. (1781/1964). *Kritik der reinen Vernunft [Critique of Pure Reason]*. N. Kemp Smith (trans.). London: Macmillan.
Locke, J. (1690/1975). *An Essay Concerning Human Understanding*. P. H. Nidditch (ed.). Oxford: Oxford University Press.
Mackie, J. (1976). *Problems from Locke*. Oxford: Oxford University Press.
Nagel, T. (1986). *The View from Nowhere*. Oxford: Oxford University Press.
Nolen-Hoeksema, S. Fredrickson, B. L., Loftus, G. R. & Wagenaar, W. A. (2009). *Atkinson & Hilgard's Introduction to Psychology*. Andover: Cengage Learning.
Nordby, H. (2004). Concept Possession and Incorrect Understanding. *Philosophical Explorations*, 1.
–. (2005). Art and Radical Interpretation. *Nordic Journal of Aesthetics*, 3.
–. (2013). *Etikk i barnevern [Ethics in Child Welfare]*. Oslo: Gyldendal akademisk.
Peacocke, C. (2008). *Truly Understood*. Oxford: Oxford University Press.
Pettit, P. (2004). The Possibility of Aesthetic Realism. In P. Lamarque & S. Olsen (eds.), *Aesthetics and the Philosophy of Art*. Oxford: Blackwell.
Pettit, P. & Jackson, F. (1995). Functionalism and Broad Dontent. In A. Pessin & S. Goldberg (eds.): *The Twin Earth Chronicles*. New York: Sharpe.
Rokka, M. (2002). René Descartes. In S. Nadler (ed.), *A Companion to Early Modern Philosophy*. Oxford: Blackwell.
Skribekk, G., Gilje, N. (1987). *Filosofihistorie [The History of Philosophy]*. Oslo: Universitetsforlaget.
Snævarr, S. (2008). *Kunstfilosofi [The Philosophy of Art]*. Bergen: Fagbokforlaget.
–. (1977). *Hume*. Oxford: Routledge.
–. (1995). Primary and Secondary qualities. In J. Danc & E. Lepore (eds.), *A Companion to Epistemology*. Oxford: Blackwell.
–. (2000). *The Quest for Reality*. Oxford: Oxford University Press.
Taylor, A. (2001). *Plato: The Man and his Work*. Mineola: Dover Books.
Wittgenstein, L. (1922). *Tractatus Logico Philosophicus*. Oxford: Blackwell.
–. (1953). *Philosophical Investigations*. Oxford: Blackwell.
–. (1966). *Lectures and Conversations on Aesthetics, Psychology and Religious Belief*. Oxford: Blackwell.
Wright, C. (1991). *Frege: Tradition and Influence*. Oxford: Blackwell.

MASS 2 – pendant – 2011 – granite, silver, nylon cord – 42 × 30 mm – private collection, Norway

MASS 1 – pendant – 2009 – granite, silver, leather cord – 60 × 45 × 28 mm – KODE Art Museums of Bergen, Norway

X MASS – pendant – 2009 – granite, silver, leather cord – 60×45×15 mm

AXIS – pendant – 2010 – ebony, silver, leather cord – 25 × 150 × 10 mm – National Museum of Decorative Arts and Design, Trondheim, Norway

HOURGLASS – pendant – 2011 – silver, glass, sand, leather cord – 125×30×19 mm – private collection, Norway

QUADRANT – pendant – 2011 – mahogany, silver, nylon cord – 85×85×5mm – private collection, Norway

APPLE – pendant – 2011 – birch wood, oil paint, leather cord – 50 × 55 mm – private collection, Norway

MOON – pendant – 2011 – granite, silver, leather cord – 55 mm – private collection, Norway

BLACK HOLE – pendant – 2011 – granite, silver, leather cord – 60 × 57 × 35 mm – private collection, Norway

MILLSTONE – pendant – 2011 – granite, silver, leather cord – 17 × 60 mm – private collection, Norway

ON IDENTITY IN ART AND LIFE
A CONVERSATION BETWEEN ANDRÉ GALI AND REINHOLD ZIEGLER

ANDRÉ GALI: In *Cosmic Debris* you have worked with meteorites both as material and as theme. I can see many reasons why you have chosen to work with meteorites, but it would be interesting to hear why you have chosen this particular material.

REINHOLD ZIEGLER: To me, meteorites are a natural variation on the basic theme in my art: using jewellery as a medium to express something about the larger, non-personal aspects of existence. For my first solo exhibition, I worked with gravity as the theme; a universal, extensive force of nature that exists independently of the individual's will and preferences. I have chosen meteorites as the theme in my second exhibition because it is the material that physically lies furthest away from us. Most meteorites come from asteroids that orbit between Mars and Jupiter. This material thus tells us something about conditions beyond life on earth.

AG: How did you get this idea specifically?

RZ: I saw a photograph of a meteorite in a newspaper and sensed that the expression was very close to an aesthetic idea in my mind. I read the article and learned that meteorites are residue from the creation of our solar system, and contain some of the proto matter from which our solar system is built. At that point I realised that this material to the highest degree coincided with what I work with, both aesthetically and thematically.

AG: You state that you are interested in the non-personal, something I take to mean a distance from the individual, but at the same time you cultivate this material, you turn it into jewellery and in so doing, it comes nearer to the individual.

RZ: Yes, and this nearness, the way I see it, is a precondition for experiencing what is distant. Distance and nearness are not necessarily opposites, but rather two terminological extremes of the theme *space between objects*.

AG: A more common strategy in art jewellery is to exploit this intimacy in order to create jewellery that serves as an identity marker, meaning a kind of token of the personality of the person wearing the jewellery. Can I understand your project as a wish to depart from this?

RZ: I believe that the role of jewellery as an expression of the wearer's personality is too one-sided in art jewellery. I can see how the jewellery's closeness to the body would make this kind of approach natural, but my claim is that there are other possibilities in the medium of jewellery that this focus on individuality conceals. I myself found another approach after a meeting with the French philosopher Georges Bataille, who pretty soon became my intellectual source of inspiration. The entire foundation of my work is actually built on the following statement from his book *Eroticism*:

'We find the state of affairs that binds us to our random and ephemeral individuality hard to bear. Along with our tormenting desire that this evanescent thing should last, there stands our obsession with a primal continuity linking us with everything that is.'[1]

What Bataille says here, is that the human being is caught in a kind of dilemma, where on the one hand it wants to defend and justify its position as a unique individual, while on the other hand it longs for the experience of being part of something bigger or something more enduring, which he sometimes calls *continuity* and other times more poetically *everything that is*. My intention is to create various expressions for this *continuity*. I attempt to treat jewellery objects as some kind of symbolic connection between the individual and *everything that is*.

AG: But is there not some kind of a paradox in your project? When you wear a piece of jewellery in public, it is bound to function as an identity marker, since people in your environment will interpret the piece of jewellery as an expression of the identity of the person wearing it. At the same time it is possible to relate to your work as mere objects, not as objects to be worn. But that would mean that the jewellery aspect somewhat diminishes.

RZ: Yes, and this paradox is actually what makes jewellery meaningful to me. My objects are made as jewellery and may be worn on the body – the fact is that I am very concerned with developing technical and aesthetical solutions that make them as usable as possible – but at the same time my opinion is that you do not have to wear them. And this is the reason why I apply the compound term *jewellery objects* to my work. I was very inspired the other day when I read a statement from Peter Skubic: 'The concept of jewellery as having a function only when being worn is something I abhor.'[2]

The way I see it, there is a great reflexive potential in relating to jewellery as mere objects. The jewellery aspect creates an intimacy to the wearer or the beholder, while the object aspect creates distance. If, applying this horizon of understanding, you still choose to wear the jewellery, it would not necessarily be as an identity marker or a social sign, but perhaps with the intention of sharing a more all-encompassing attitude with others – something which happens to be my motivation for working with jewellery objects.

AG: It would be interesting, in light of this, to talk about how you developed this idea that the piece of jewellery does not necessarily have to be an identity marker. You have studied jewellery, and one could have expected a more traditional approach.

RZ: The turning point occurred on a study trip to Greenland in 2003. At the local museum in Nuuk, I came across a series of talismans; small leather bags filled with magical material. They were placed on the inside of the clothes with thin leather straps. The fact that these objects were hidden from view greatly fascinated me. At the same time I became practically obsessed with a desire to find a rational approach to the superstition in this tradition. My first angle of approach was to view the talisman as an expression of the irrational.

AG: What you say here about the rational and irrational leads me to think of Friedrich Nietzsche, who referred to these terms respectively as the Apollonian and the Dionysian forces. You mentioned that you are inspired by the French philosopher Georges Bataille. His thoughts are very much based on Nietzsche's, the way I see it. At least they share the notion that the irrational has been lost in Western culture, and that this has happened at the cost of something deeper within the human being as an individual. Bataille's project is really about a search for something primal, something not contaminated by a modern rationality, and he looks to primitive communities and ancient religions for answers. I think it would be interesting if you could shed some light on this in relation to what you do.

RZ: Yes, this is important to me. What I believe lies in primitive communities, and that has disappeared in our rationalized culture, is what Bataille speaks of as *transgression*. By that he means that there are activities that remove individuals from themselves, and the position they hold in everyday life, which Bataille incidentally calls *work*. An important part of his project is to consider the sexual drive as a natural part of the same phenomenon. The sexual act is about transcending yourself as an individual. In fact he has often used the word 'eroticism' as a synonym for *transgression*. This insight became my second angle of approach in order to understand what the talisman phenomenon is all about. When I removed the magical element I found that the remaining function of the talisman was to make the user give up his individuality and to connect to *everything that is*.

AG: What you say about *transgression* makes me think about how rationality is linked to control – more specifically, how the individual tries to control its own image outwards, that is how one presents oneself to others. Jewellery objects as identity markers are thus part of this control mechanism. You state that in the *transgression,* this control is abandoned.

RZ: Yes, it is this understanding of identity, which is about constructing an image of yourself in other people's consciousness, that I want to introduce an alternative to.

AG: You say that Bataille is a great source of inspiration to you. For a long time I have been interested in another French thinker, Jean Baudrillard, who bases some of his ideas on Bataille, and who has said a lot about structuring this image of the self. In the 1960s Baudrillard was concerned with analysing Western consumer society. One of the points he makes is that consumer society profits from the fact that we never feel whole, or that we need something external in order to *fully become ourselves*. What we lack we can obtain through buying products. In consumer society, the way Baudrillard understands it, everything you buy – or consume – is a sign of personality. He reads social reality as consisting of signs, put together in various sign systems that express something of who we are and where we belong in society – and in the social hierarchy – meaning identity. For example, the car you drive might say something about who you are. For someone climbing the social ladder it might be important to show signs of wealth by buying a car that everybody knows is expensive, or that is large and flashy and in this way signals strength. However, for someone born into a cultural and economic elite, it might be a point to under-communicate these values. Then you have to be part of the same social community in order to understand what the car really communicates.

RZ: Either way this identity will be a construction and will not necessarily say anything about who we truly are. As a matter of fact, the opposite is more likely; our attempts to structure our own self-image stops us from seeing who we truly are. I have always believed that we all have a deeper base or an essence within us, where our true nature lies. To find out more about this, I started meditating a few years ago, starting with various breathing techniques and moving on to a technique known as the Ishayas' Ascension. This old Eastern tradition dates back several thousand years, and it was introduced to the Western world about 20 years ago. Central to this tradition is the belief that we possess an essence, and that we have to transcend our constructed self-image in order to experience our essence. For me, then, this meditation has become a concrete means to experience *transgression*.

AG: Hearing you say this, I want to pursue Baudrillard's thoughts a little further. If we look at the art world in the light of his perspective on consumer society, it is clear that one aspect of art is that it functions as some kind of identity marker, showing one kind of cultural competence. That you are familiar with, and buy one type of art, says something about your values, your competence, your interests and so on, and it connects you to other people with similar values and interests. My thought is that when jewellery wants to enter the art world – when it claims to be something more than "simply" jewellery – the jewellery still functions as a consumer product, but in another value system. By buying a piece of jewellery worth 4,000 euros, – made by say Peter Skubic, you signal something other than wearing jewellery bought at H&M. It shows that you are special, that you are conscious about what you wear, that you are familiar with jewellery artists. The fact that there is a history of creation behind it, one you won't find in jewellery from H&M, is also important. The craftsmanship, the conceptual quality and the fact that it is made by an artist – these things in fact make art jewellery more individualizing than mass produced jewellery. This is a paradox. Within this field, then, your position is also paradoxical, not least because you make unique objects.

RZ: One response to that is that the work of art itself has a life beyond having to be owned by someone, for instance as an exhibition object at a museum, as an image or simply something talked about. Then it functions more as an object we experience, freed from its role as an identity marker.

AG: But experiences can also be used as identity markers. Seeing the works of an artist we really like at a museum or in a book, and by sharing this experience with others, for instance by posting it on Facebook or Instagram, we make a statement of our personal taste.

RZ: The second response to this problem is that it depends on the intention with which you approach art. Art can exist as merchandise, and art can exist as an experience – even as a spiritual experience. The way I see it, art is a tool we can use in order to cultivate our ability to experience the world. And this is precisely what art has in common with several spiritual movements, especially those based on meditation. Most meditative techniques, including the Ishayas' Ascension, are forms of awareness training, where the goal is to be able to observe without falling into the temptation of judging or analysing. The more aware we are, the clearer we will see that the constant stream of personal 'comments' on reality is based on a distorted image of ourselves and the world. This self-image is usually based on a mixture of previous experiences and wishes for the future. And this hinders us from living here and now – which, according to the Ishayas and many other spiritual movements, is the only state in which we can experience ourselves and the world as we truly are. True awareness can only be reached by transcending the constructed self-image we talked about earlier. I have practised this technique for four years and can safely say that my way of experiencing is much richer now, and my art is a result of this richer experience. Having said this, I have to emphasise that it is important to me that my works be able to function as independent objects, without my additional explanations. It must be up to the audience to decide how they wish to experience a jewellery object – as an identity marker or as an object that can stimulate awareness.

AG: It is of course possible to do both. Bataille talks about *transgression* and *work* replacing each other successively.

RZ: The answer to that is both yes and no. In a spiritual understanding, it is not possible to combine a wilful understanding of identity with living here and now. As soon as there is an intention of achieving something, you have lost the innocent awareness of the present, and the experience is disturbed. This is very subtle and might not be understood until you have experienced how consciousness functions through meditation. When that is said, I have to add that living entirely in the present moment demands a high level of consciousness, and that for most people, this only exists as an ideal. Before having reached this higher level of consciousness, there will always be a successive interaction between the two perceptions of reality, but the ideal is to be able to experience life even more in the here and now.

AG: I would like to move our conversation back to a more concrete level. You have made a series of jewellery objects where the pieces of jewellery look like tools, for example hammers with granite heads, and thongs made in oxidised silver. Do these works come under the same theme?

RZ: The functional aspect is meant as a vehicle to lead attention away from the purely decorative aspect, which is so closely associated with the notion of jewellery. At the same time, I try – by transcending the original function – to create associations with another kind of function, for instance a symbolic or ritual act. But I have not wanted to say anything about which act is being talked about. The user or onlooker must imagine that himself. The central issue here is that a ritual is – according to Bataille – an act where you transcend yourself.

AG: These works give me certain associations both to freemasonry and to the objects we find in Matthew Barney's art. When I spoke to you about art jewellery in connection with the exhibition and the book *Aftermath*, we spoke about your sources of inspiration, and you told me that you are inspired by artists such as Barney, so this association is perhaps not coincidental. Or is it?

RZ: The kinship to Barney is not coincidental. He is probably the most important source of inspiration when it comes to my *Tools*-series, apart from Bataille of course. But the association to freemasonry is more an indirect consequence of the fact that Barney plays on rituals from the freemasons in some of his work. I am not a freemason, nor am I particularly interested in it, apart from the fact that they might teach me something about rituals. I have been told in retrospect that the hammer plays a certain role for Freemasons. To me it has been important that the hammer is an ancient symbol found in many cultures and religions, for example in Buddhism and in Norse mythology.

AG: Many of your works also have a mechanical part that the user can set in motion. Is that meant to be an invitation to this type of ritual act?

RZ: Yes, they all offer the opportunity to perform an action, but the action is not functional in a traditional sense. In addition, it is important that they have a technical expression that can support the notion of being connected to something.

AG: And what about the brooches that have been shaped like breast pockets, do they also have something to do with *transgression*?

RZ: The pocket brooches are a play on functionality. You can keep something practical in them, or something of symbolic meaning. Some might want to use them as talismans. At the same time, the pockets are meant to play on the jewellery's relation to clothing, since they initially look like a part of a person's clothing. The brooches do not really claim to be jewellery at all, and are thus not well suited to function as identity markers.

AG: So far we have talked about the functional aspects in your work. Another aspect that is quite striking is the inclusion of nature, perhaps especially since you, in almost all your work, use traditional natural materials like leather, wood and stone. In your own essay *Gravity*[3], accompanying the exhibition with the same title, you stated that the element of nature in your art not only has to do with aesthetic beauty, but that it is a natural consequence of Bataille's influence on your art. Could you explain this further?

RZ: To Bataille, the rationality that constitutes life in *work* is a 'no' to nature. We organise our communities to control nature, both its destructive and its productive powers. *Transgression* is in turn a 'no' to *work* and a return to nature. So, my approach to nature is using it as an expression of *transgression*.

AG: In art jewellery there is a clear consciousness about the wearer and the wearer's body. As you are so concerned with departing from jewellery as an identity marker, and even advocate that jewellery does not have to be worn at all, how do you relate to the body in your art?

RZ: I view the body as an amazing vehicle that enables us to experience; through a multitude of sensory organs. Even thoughts and feelings take place in the body. It is easy to forget this when you treat the body as a showcase of the individual's personality, as we find in various body cultures and in many kinds of art jewellery. Everyone is unique, and everyone has a unique body, but we *are not* our bodies.

AG: It has become very popular to say that the body is the interface, or the meeting point between the individual and reality. You can physically experience reality, or you can think of it in terms of concepts. The intellect is the starting point for conceptuality in art. What role does it play in your work?

RZ: It's easy to assume that our thoughts can express who we are. As you mentioned, rationality can be traced back to the Greeks and maybe further back, but in our Western world I suppose it is Descartes who is considered the father of rationality in that respect. I think, therefore I am!

This has taken up such a great part of our culture that it's easy for us to believe that we *are* our thoughts. But in a spiritual understanding, we are not our thoughts. We do have thoughts, but they do not express who we are. Many of the techniques within the Ishayas' Ascension are about distancing yourself from your thoughts, not stopping them completely, but to understand that they are just thoughts that come and go.

AG: What role do feelings have in this?

RZ: The same applies here as well, we are not our feelings. I see that many of the artists that treat the jewellery medium as an identity marker have a tendency to become overly sentimental, for instance by focussing on memories and affective values. This sentimentality also lies behind the material-based expressionism we have seen in the last few years, where materials and colours are mixed with each other in collage-like ways, and where difficult craft techniques are abandoned, so that the jewellery, to a large extent, becomes a result of the artist's intuitive and subjective whims. There are, however, some artists that work with feelings in a convincing way, but a large part of it is pretty insignificant. In addition, I claim that a one-sided worshipping of sentimentality is a delusion. We have feelings, and they often provide important clues to life, but we *are not* our feelings.

AG: This thing about who we really are I find interesting. If we follow Baudrillard, there is no *self* to speak of, not even an objective reality. Everything is signs, formulated and reformulated in an on-going flow – he calls it *Simulacrum* in one of his best known books, a book that by the way was an important source of inspiration for the film *Matrix*. It also points to Descartes and what you say about rationality establishing what is real. But it isn't this simple to Baudrillard, because he believes that some kind of a personal essence, a self, or even an objective reality quite simply does not exist. Quite the opposite, he believes that the world's purpose, and especially the visual world's, is to cover up the fact that there is nothing beyond. Everything is staged, even the self. So, considering this, I am curious as to what you believe our essence is.

RZ: I disagree with Baudrillard on this. We have a stable and peaceful *place* inside us, which is a completely natural state most people have experienced once or many times in the course of life. The easiest explanation is to say that this *place* is our pure consciousness, a silent presence where thoughts cease to exist. The American psychologist Abraham Maslow called it *peak experiences,* and described it as a kind of flow zone where things happen without effort. He referred for instance to artists who are immensely creative, athletes who are capable of performing something extra and women who give birth. To him it was important to demystify this phenomenon and describe it as a common experience. The good thing about meditation is that we learn techniques that we can use to enter this state whenever we want to. It can even become permanent. The philosophy of the Ishayas' Ascension is that as you eventually, by refining awareness, get better connected to your pure consciousness, your natural *self* will present itself automatically and without effort. I myself have not yet experienced this fully, so I can't prove that *this is how it is,* but my experiences so far indicate that this is true.

AG: Since you to such a degree go against the most dominant movements within art jewellery, how do you relate to your own field, for instance, do you have any role models?

RZ: I pay close attention to what goes on in the field and I think there are many artists who make both good and interesting works. I have already mentioned Peter Skubic. His retrospective exhibition *Radical* in the Pinakothek der Moderne in Munich in 2011 was magnificent. Bruno Martinazzi, with his interest in physical distance and fragmented human bodies, creates works that I find both interesting and beautiful. However, the jewellery artists that I feel most closely related to are Dorothea Prühl and Warwick Freeman, perhaps particularly due to their significant use of nature; Prühl through her motifs and material choices, Freeman through his influence from the indigenous population of New Zealand, the Maoris.

AG: We have now spoken a bit about the tradition and where you place yourself within it. In this context it would be interesting to talk some more about the *Aftermath*-project, which you were part of. You were eight very different artists, some at the start of their careers, others more established. I spoke to all of you for the book that was made in connection with the exhibition. The Munich-based artists have all studied with Otto Künzli. He has been a dominant figure in the history of art jewellery, both as an artist and a teacher. I feel that this was a project that reflected on some of the history of art jewellery, but at the same time wanted to look forward, something the title refers to.

RZ: The artists for *Aftermath* were picked because they had different strategies as to what could be fruitful in the future. The common denominator for the *Aftermath* project is that the chosen artists challenge the term jewellery in some way, so that it is not all about decoration. For those who are familiar with art jewellery, it is perhaps obvious that jewellery is more than adornment, but as soon as you step outside the informed circle, the situation is different. Here the notion of jewellery is closely connected to feminine decorative values. So it is a term that is problematic and very hard to deal with.

AG: We discuss this jewellery term in the *Aftermath* book, and there have been different views as to where art jewellery belongs, and whether it is art at all. I believe there are two solutions: that art jewellery is part of a larger field of art, or that it constitutes a field of its own, independently of others. A challenge within the field of art jewellery is that it is quite a small field, but despite being so small, a lot of things are happening, though the knowledge of this might be limited to especially interested people. It seems to me that the fact that you had the exhibition at the Vigeland Museum in Oslo and Museum Villa Stuck in Munich, and not in jewellery galleries or industrial art museums, is making quite a statement.

RZ: Yes, it was definitely very important that this wouldn't become a jewellery project about jewellery for jewellery 'people', which is a strong tendency within the field. However, I see that this internalisation has been necessary so that the field could evolve and find out what might lie beyond the goldsmith tradition. Among other things, we have seen many artists making jewellery about jewellery. This has been important when it comes to the field's understanding of itself, but I believe we should be past that now. It is about time we try to communicate wider. Jewellery may not be considered art to the same extent as fine art, but part of art jewellery might get some attention in a broadened field. I really hope so.

AG: I have been interested in the way jewellery is exhibited. It is often shown in museums as objects, and more likely than not in showcases. In the *Aftermath* conversations about what art jewellery is many state that what is special, is that you can wear it. But when it is shown as an object in a museum in a connection with art, it is the sculptural and aesthetic qualities that are emphasised.

RZ: That is an interesting observation. I have noticed that *Aftermath* has been criticized for having exhibited in aquarium-like showcases, and that this limits the further life of the works and that it would have been better having presented them openly, so that people could touch them, and perhaps try them on. In my opinion, what happens when you exhibit jewellery in showcases is that the object aspect is emphasised and the decoration aspect diminishes. And this was a completely conscious strategy in the *Aftermath* exhibition.

AG: In closing I want to ask you about the title. *Aftermath* refers to The Rolling Stones and their breakthrough album from 1966. The title of your project, *Cosmic Debris,* leads me to think of Frank Zappa's song with the same title from 1974. It's about a guru, "The Mystery Man", offering Zappa nirvana for a small sum of money. Zappa reverses the situation and hypnotizes the guru and steals his belongings. It is a song that in my opinion mocks the kind of things you talk about; spirituality, rituals and meditation. He says: 'So take your meditations an' your preparations An' ram it up yer snout.'[4] Why have you chosen this title, and what does it mean to you?

RZ: Well, I am a Zappa fan and think it's a really cool song. Having said this: The literal meaning of the term *Cosmic Debris coincides with what meteorites really are – residue from space*. It is also the title of a well-known book on the subject. In other words, it is quite appropriate, and at the same time as it 'snobs down' instead of up. It actually took me a long time to find a term that could describe the meteorite theme, because most terms I came across had too strong associations to dubious alternative movements. Even though I am interested in spirituality, I still have reservations when it comes to things that are out of touch with reality and quack movements, including several religions. I think it's a good thing that Zappa's song mocks this. I'm happy to let Zappa keep me on my toes. It helps to balance my project. It reminds me that no matter what my intentions are, I can't keep my audience from doing like Zappa: Running off with my work and telling me to take my *meditations and preparations and ram them up my snout*.

NOTES
1   Bataille 2012, p. 15.
2   Cherry 2013, p. 104.
3   www.ziegler.no
4   www.elyrics.net, Cosmik Debris. Accessed 20 Oct. 2013.

REFERENCES:
Bataille, G. (2012). *Eroticism*. London: Penguin.
Cherry, N. (2013). *Jewelry Design and Development*. London: Bloomsbury.
www.elyrics.net, *Cosmik Debris*. Accessed 20 Oct. 2013.

CROSS PEEN HAMMER – pendant – 2013 – granite, ziricote wood, silver, nylon band – 155×70×21mm

PINCER PLIERS – pendant – 2012 – silver, leather, nylon band – 150 × 42 × 20 mm – private collection, Germany

ORANGE POUCH – pendant – 2012 – leather, silver, nylon zipper, nylon cord – 85×40 mm

PLUMBERS PLIERS – pendant – 2013 – silver, nylon cord – 150 × 32 × 10 mm

HAMMER – pendant – 2012 – granite, mutyenye wood, silver, nylon band – 150×65×22 mm – private collection, Germany

M A S K – mask – 2012 – leather, silver, nylon band – 200 × 50 × 4 mm

VIOLET POCKET – brooch – 2013 – leather, silver, brass, steel – 95×70×10mm

REINHOLD ZIEGLER

1965	Born in Kristiansund, Norway

EDUCATION
2001–06	The National College of Art and Design, Oslo, Norway
1989–90	Jewellery Design, Staatliche Zeichenakademie Hanau, Germany
1984–87	Goldsmithing, Elvebakken Polytechnic High School, Norway

SOLO EXHIBITIONS
2014	*Cosmic Debris*, Galerie Wittenbrink FünfHöfe, Munich, Germany
	Cosmic Debris, Kunstnerforbundet, Oslo, Norway
2011	*Gravity*, Galleri Format, Oslo, Norway

GROUP EXHIBITIONS
2013	*Révélations*, Grand Palais, Paris, France
	Aftermath of Art Jewellery, Museum Villa Stuck, Munich, Germany
	Aftermath of Art Jewellery, Vigelandsmuseet, Oslo, Norway
2012	*Kunsthåndverk 2012*, The National Museum of Art, Architecture and Design, Oslo, Norway
2008	New and Norwegian – *A World of Folk*, Stavanger, Norway
2007	*European Jewellery Exhibition*, Østfold Kunstnersenter, Fredrikstad, Norway
2006	*Nine Materials*, Galleri Platina, Stockholm, Sweden
2004	*Triennale*, Oslo, Norway/Leipzig, Germany

AWARDS
2005	3rd Prize Norwegian National Design Competition, Souvenirs

PUBLIC COMMISSIONS
2010	Major Chain for Modalen, Norway
	Major Chain for Iveland, Norway
2007	Major Chain for Aremark, Norway

GRANTS
2012	Three-year working grant, Norwegian Government
2010	General grant, Norwegian Government
2009	Study grant, Norwegian Government
2008	Establishing grant, Norwegian Government

ACKNOWLEDGEMENTS

I would like to thank the following for their contributions to this project:

Philosopher Halvor Norby for opening up new horizons in my working process and for writing a wonderful essay. André Gali for his stimulating discussions and profound insights. Arlyne Moi and Cora Dannatt for their patience and tenacity in translating. And Emmanuel Aquino for helping me find precise titles for my works.

Marion Boschka and Dirk Allgaier at Arnoldsche Art Publishers, designer Silke Nalbach and colour separator Rüdiger Mayer for creating the perfect book for my art.

Morten Bilet for his supply of meteorite material and knowledge.

Eirik Selvaldsen for his enthusiasm and for letting me access his enormous personal network.

And last but not the least my studio colleagues: Sigurd Bronger for his unique combination of infectious enthusiasm and constructive criticism, and for his wonderful photos of my meteorite works; and Runa Vethal Stølen for her interesting dialogues and for being ever so positive even in the midst of challenging circumstances. Your support has been essential in bringing this project into fruition.

Text Norwegian